George Hinckley Lyman

The Interests of the Public and the Medical Profession

The Annual Discourse before the Massachusetts Medical Society, June 9,

1875

George Hinckley Lyman

The Interests of the Public and the Medical Profession
The Annual Discourse before the Massachusetts Medical Society, June 9, 1875

ISBN/EAN: 9783337816414

Printed in Europe, USA, Canada, Australia, Japan

Cover: Foto ©berggeist007 / pixelio.de

More available books at **www.hansebooks.com**

THE INTERESTS OF THE PUBLIC AND THE
MEDICAL PROFESSION.

THE

ANNUAL DISCOURSE

BEFORE THE

MASSACHUSETTS MEDICAL SOCIETY,

JUNE 9, 1875.

BY

GEORGE H. LYMAN, M.D.

BOSTON, MASS.

BOSTON:
PRESS OF DAVID CLAPP & SON.
1875.

DISCOURSE.*

Mr. President and Fellows

of the Massachusetts Medical Society :—

The first annual reunion of our Society, comprising at that time but thirty-one members, was held ninety-four years ago.[1] The custom inaugurated three years later, of devoting an hour at these meetings to an annual address, has, with four or five omissions only, been continued now for nearly a century. From the mere handful of men originally composing the Society, we have increased to some fourteen hundred members.

During this long interval, the changes which have come over the spirit and practice of the profession of medicine, and the altered relations developed between it and the public, are more or

---

* At an Adjourned Meeting of the Mass. Medical Society, held Oct. 3, 1860, it was

Resolved, "That the Massachusetts Medical Society hereby declares that it does not consider itself as having endorsed or censured the opinions in former published Annual Discourses, nor will it hold itself responsible for any opinions or sentiments advanced in any future similar discourses."

Resolved, "That the Committee on Publications be directed to print a statement to that effect at the commencement of each Annual Discourse which may hereafter be published."

[1] November, 1781.

1

less familiar to you, as are also the persistent attempts of unqualified pretenders to get such a lodgment under our banner as would enable them the better to pursue their impositions. Occasionally unfit persons have by false pretences obtained admission, but the association has proved an uneasy one for them, so incongruous and uncongenial, that sooner or later they have been compelled to retire.

The history of these innovations, but more especially some detailed account of the public work effected by our organization, and of the changes which higher attainments in knowledge, resulting from the rapid discoveries effected in physiology, pathology, chemistry and microscopy, have required in our practice, would be an instructive and entertaining subject for our rapidly approaching Centennial. I can merely allude to them incidentally, premising the assertion that we may honestly claim that all these controversies as to ethics, membership, qualifications, etc., have resulted to the advantage not merely of ourselves, but of the community, for whose best welfare alone we have any claim to exist.

Were our objects purely selfish we might well abandon every struggle for better things, for who does not know that a patent pill or a well advertised elixir has more money in it than our philosophy has ever dreamed of ?

If we are accused of conservatism, it must be conceded that it is a conservatism alone of the interests of the public; if we have refused to affiliate with any " exclusive dogma," whether it be Perkinsism or Spiritualism, Thomsonianism or Eclecticism, Allopathy or Hydropathy, Homœopathy or Electropathy, our sufficient reason is that the range of the regular practitioner includes everything of worth which they contain, most of these erratic systems originating mainly in attempts to erect an independent structure upon some isolated but already well known medical fact, valuable only in its existing and subordinate connection.

The history of medicine for a thousand or more years is indeed but a history of successive grafts upon the main trunk, of original theories and discoveries, whenever careful investigation has shown a reasonable probability of their ultimate fruitfulness.

This Society has always been as eager to welcome any new idea — proved to be worthy its welcome, as it has been wisely sceptical in regard to any novelty intruding itself without satisfactory credentials, believing that the true spirit of scientific research and the only safe method in these days of modern culture, each giving birth to some fancied new discovery, is carefully and patiently to analyze them all, that rejecting the chaff it may give its sanction only to that which practical experience proves to be good.

We place no limit upon this spirit of research. Every investigation of disease, every experiment with drugs, every physiological problem which may properly be pursued outside our Society, may quite as well, if not better, be done within it; certainly there is no restriction upon any one's liberty of action, other than the check which an association with others engaged in similar pursuits will always impose upon rash procedures or hasty deduction.

The record of the Massachusetts Medical Society shows that from its origin continuous efforts have been made to raise the standard of preliminary education,[1] to encourage the development of every scientific discovery in the various branches of our profession, and by rigid observation of their practical applications, to guide the suffering through their unavoidable attacks, relieving those organic disorders which are incurable, and rectifying, so far as may be, those functional derangements which impede the healthy processes of nature.

Were the charge that our organization tends to limit free inquiry and independent practice, true, we should now be all following the same routine of set formulas for every disease. So far from

---

[1] Within a few weeks the Harvard Medical School has announced a long-needed step in advance. After September, 1877, students who desire its advantages must possess a degree in Letters or Science from some recognized College or Scientific School, or else pass an examination in Latin and Physics. The By-Laws of the Massachusetts Medical Society have always professedly required these qualifications for admission, but from the necessities of the case the rule has been practically ignored. This initiation of a better state of things by the Medical School comes none too soon. It will meet with the cordial approval of the Society, and it is to be hoped that the example will be followed elsewhere. In earlier days the facilities for education were so limited that a reasonable excuse existed for laxity on the part of the schools, but the time for such exigencies has passed, and it is now no longer necessary to matriculate indifferently prepared men. The higher the standard the more attractive will the profession become to the class of minds needed for the rapidly increasing scientific developments of the day.

the truth is this, that the disagreements of doctors
are proverbial.

While pathology, physiology, diagnosis, hy-
giene, the natural history of disease, have in their
rapid development approached more and more
nearly the dignity of science, it still remains true,
that the practical application of our remedial
measures to the multiform and complicated emergencies arising from individual constitution and
habits of life, constitutes as yet only an art, in
which the artizans, precisely as in all other callings, are more or less skilful, and that the fullest
latitude is allowed to the judgment of each individual. Every physician has his favorite method,
the result of personal experience or choice, for
meeting indications as they occur. One is most
successful with morphine, another with opium as
an anodyne; one prefers cold water, another,
digitalis and squills, for diuresis; one approves
of blue mass, another of podophyllin; one believes
in the ligature, another in torsion. Some, indeed,
believe in nothing at all, unless it be with the
early objectors to anæsthesia in midwifery, that
pain and suffering are a part of the plan of Crea-
tion and had better not be much meddled with!

It cannot be denied that there is quite as much
difference among regular practitioners in their
methods of treating the same disease as between
them and the better class of so-called irregulars,
between whom and ourselves the distinction in
many cases is one of ethics rather than therapeutics, and the result has naturally been the ab-

sorption by regular practice of all worth preserving in every ism and pathy that has attempted the erection of a new school on its " exclusive dogma." Nor need we fear a different result from any of the novelties which the future may have in store.

The restless spirit of the age resents any exhibition of conservatism, no matter where it may manifest itself; neither law, theology nor politics is safe from the attacks of the modern reformer, clad in his mail of self-conceit. It is not surprising, therefore, that our own profession, a compound as it is of science and art, and which is perhaps the least capable of mathematical demonstration of any of the liberal callings, should be obliged, in its turn, to put itself on the defensive against outside clamor, nor that the cautious spirit which has always characterized our action as a Society, has been attributed, by the unthinking portion of the community, to illiberality, to a blind conceit and pride of opinion, clouding our vision of the new truths so palpable to their clearer intellectual insight.

To evince their contempt for our stupid obstinacy, such critics rush headlong to the opposite extreme, and in the unseemly haste to exert their influence in favor of every untried wonderful new revelation, forget that none are so ready to recognize a new truth as those who after "making haste slowly" have given themselves time to test its truth, and that the judgment of a body of well-trained experts is more reliable than the crude fancies of enthusiasm and one-sided experiment, or the bold and reckless trading upon the prejudices of the masses, so characteristic of many of the reformatory movements of the day.

It surely may, without presumption, be claimed for regular practitioners as a body, that they are as intelligent, honest, unselfish and gifted with as fair a share of common sense, as any other body of men. Granting this, it is passing strange that the same business sagacity which distinguishes our people in their respective callings, should not lead them to recognize that their safest course, in matters involving such vital issues as the preservation of life and health, is to leave the working out of these problems to those specially trained for the duty, and whose interest in the result cannot possibly differ from their own.

The increase of this restless spirit, in these latter years, has gradually effected marked changes in the relations of all professional men to the public. The physician, especially, instead of being looked up to with deference and respect as the nearest and dearest friend, is only too often regarded as a mere mechanician, whose services are to be cheapened if indeed they are paid for at all, and whose relations to the family are considered much on a par with those of the butcher or baker who supplies their daily food. I do not mean to assert that this is always or even in the majority of instances true, for we all know how affectionately we are relied upon by very many of those to whom we minister; but I do mean to say that the spirit of quackery has tended greatly to diminish that

kindly cordiality and confidence which was so po-
tent as an incentive, and that it has induced many
of those who, nominally belonging to us, cannot
be said to be really of us, so to lower their standard in deference to this state of things, as to become little better than the money-getting empirics,
and furthermore that the encouragement of this
spirit by the public has wrought much detriment
to their own interests.

The question naturally arises, are the public
wholly to blame in this matter, and may it not be
that we have been too sensitive in our estimate of
their criticism? We cannot deny their right to the
exercise of an independent judgment upon the
merits of any new system proposed for their
acceptance. Neither ridicule nor denunciation
will have any other effect than to strengthen their
sympathies for those who profess to be only striving
after the truth. Truth must always come to the
surface sooner or later, and the regular practitioner
may well await the issue patiently.

The public too have found it difficult to divest
themselves of the old-time, well-founded association between physicians and nauseous drugs.
When they fully realize, as they are rapidly doing,
that modern medicine professes—not the cure of
disease, that being the province of the Almighty—
but first and foremost its prevention by hygienic
and sanitary means; and secondly the rectifying
of deranged functions, chiefly by diet, rest, tem-
perature and nursing; and lastly the relief of pain
and distress, induced by organic changes which

sooner or later must come to all as the prelude to
the end, burdens which can only be mitigated never
cured, and that in all this work drugs play but a
secondary, though still to some extent an indispensable part, the sooner will the old confidence
return.

I say an indispensable part, for with all proper
deference to the statements of such representative
men as Sir John Forbes, Sir James Clark, of
Bigelow, Holmes and Gould, there nevertheless is
danger of our placing too narrow limits upon the
usefulness of drugs. We should at least be
cautious of putting ourselves in a false position
with the public, by overstepping the point at
which the truth rests. The constant and useless
repetition of their opinions without the qualifying
statements accompanying them, has given rise in
many quarters to erroneous ideas of their real
meaning.[1]

It is unnecessary to state to this audience that
the practice of to-day is in almost every respect
unlike the practice of our fathers, and probably another half-century will witness a corres-
ponding amount of light thrown upon much
that is now confessedly obscure. Such progress
can only be delayed by an arrogant assumption
that all novelties are necessarily worthless.

Intemperate opposition only makes martyrs of
those, than whom none know better how much for

---

[1] " With the medical profession the true orthodoxy is an unquestioning
allegiance,—not to convictions and opinions held as beliefs,—but to the
duty of inquiring whether convictions and opinions correspond to the
order of nature."—*Brit. & For. Med.-Chi. Review*, Jan., 1875, p. 48.

their interest is this species of martyrdom. There is a force behind which irresistibly impels to progress and development, and it becomes the medical profession to retain the lead if they do not wish to be lost sight of in the rear. Were the axiom always borne in mind that we have no rights separable from the interests of the public, it would clear away much sensitiveness and misapprehension,—then the public realizing our disinterestedness would more cordially rely upon our judgment, and patiently wait for all necessary investigation with something of the old faith in the final decision.[1]

A marked feature in the profession of to-day is the tendency to carry the subdivisions of medical work to an extreme degree. Formerly the only division considered necessary was between medicine and surgery, and that only in the larger cities. Soon dentistry became a specialty, and properly so. Then followed the diseases of the eye and ear, and both profession and public recognized its propriety and convenience, if not necessity, until at last we have the special lithotomist, the special tenotomist and chiropodist; one man devotes himself to the perinæum, another to the palate; one confines himself to the lungs, another to the larynx; one to the brain, another to the rectum; one to obstetrics, another to uterine dis-

---

[1] A gratifying and encouraging indication of the confidence still existing in the purposes and practices of the medical profession, notwithstanding the plausible and aggressive character of quackery, is shown by the liberal contributions made since our last meeting, for the erection of new buildings to facilitate the work of the Medical Department of Harvard University.

ease; one to electricity, another to the massage, and so on — until soon each family will require as many doctors as there are different deranged organs or functions needing attention, none having that especial family interest in their patients which formerly existed, nor caring much, perhaps, beyond the fee. Some few specialties are universally recognized as both convenient and appropriate, and we may well acknowledge our indebtedness for some of the most brilliant discoveries of the century to individuals who have given especial devotion to some one branch of their profession; but we must not forget that they have been men whose general acquirements were not limited to the subjects they have illustrated, but that these were the solid foundation on which, alone, they were enabled to push their special investigations to such brilliant results.

The excessive subdivisions now in fashion are not only useless but positively injurious, and if this process of attenuation is to go on, we shall soon arrive at that degree of globular dilution which leaves nothing of original strength.

They are unnecessary, for the reason that every well-informed medical man is or should be competent for the management of all ordinary diseases; and injurious, for the reason that the general public, especially outside our large cities, are influenced thereby to lose their confidence in their own local medical adviser, upon whom they ought to be able to rely, while he in his turn, from the very prevalence of this feeling, finds no encouragement to keep himself well informed. It is objectionable also from its inevitable tendency to open up small fields to the range of superficial men, too ignorant or indolent to embrace the larger sphere of general practice. Diagnosis, pathology and therapeutics, in their wider range, are essential to the proper treatment of any special disease, medical or surgical, and he only is deserving of trust who from study and experience has gained some insight into their practical applications to all diseases.

The best operator is not always the best surgeon, and so of the rest. It is a question with many observant members of this Society whether there should not be some stipulation requiring of those now looking forward to their degrees and to admission to fellowship, and who are professedly devoting themselves to some anticipated specialty, a preliminary course of hospital or dispensary practice. Every young practitioner finds himself, after a few years of general experience, more especially fitted by taste or acquirement for some particular department of practice in which he feels himself stronger, develops more interest, or finds more to his taste. Such preferences would thus be but a natural outgrowth from experience, and it would seem to be the most suitable way of acquiring reputation for especial skill in any particular department.

There is one topic relating to our profession which is now interesting certain portions of the public, and one which, sooner or later, we may be compelled to take into consideration, arising out of the excitement as to " Woman's Rights," especially in relation to a higher standard of female education. I refer of course to the questions as to whether our medical schools and hospitals should be opened to women; if so, whether by the method of separate or mixed instruction; and finally, whether those able to pass satisfactory examinations shall be admitted to our Society. This opens up as a necessary preliminary the whole subject of woman's moral, mental and physical capacity and fitness, involving too wide a range of inquiry for the time allotted me, even had I the disposition or qualifications necessary for the work ; but the subject seems of sufficient interest to justify the appropriation of a part of the hour to some general considerations, with only incidental and unavoidable allusions to the deeper physiological questions involved, and which have been so freely discussed elsewhere.

First of all, then, we may freely admit the right of women to every educational advantage in all the Arts and Sciences, in Law, Theology and Medicine. They not only have this right, but they already exercise it; and if they can attain to the standard required, we need seek no occasion to throw obstacles in their way.

If the world has reached its present maturity and they have not as yet proved themselves, with brilliant but rare exceptions, able to compete in certain directions with their male companions, we may even willingly accept the claim made for them that they have not had a fair chance, and throw the lists freely open to as many as feel it their vocation to enter.

In many things they need enlightenment, especially in sanitary, hygienic and physiological subjects, both for their own governance and for the proper training and development of their children.

Whether the education they seek shall be in our companionship, whether the Law Schools, the Theological Seminaries, or, which more concerns us, the Medical Colleges and dissection rooms, shall be thrown open to them indiscriminately, and whether we, especially, are called upon to assist them in securing these demands and then admitting them to this Society which was never intended for such affiliation, are questions involving many other considerations by no means so clear.

Whatever may be the ultimate decision, the subject deserves dispassionate consideration, and the result must apply to the sex as a whole, and not to a few exceptionally masculine women only, whose interests and fancied rights are as nothing if they should be found to clash with the welfare of the whole body politic.

The only occasions on which this Society has been called upon for an expression of opinion are the following:—in February, 1853, the advice of the Councillors was asked by the Censors as to the application of Mrs. N. E. Clark for admission. On motion of Dr. Jacob Bigelow, it was voted that the Censors be instructed to examine male candidates only. At the annual meeting in 1867, a communication was received from the Trustees of the Massachusetts General Hospital, asking for an expression of opinion as to the expediency of admitting females as students to the wards of the hospital, and it was voted to be "inexpedient to admit females as students to our State Medical Schools and Hospitals." In October, 1872, the Councillors were asked by the Censors for instructions how to act on the application of Miss Susan Dimmock for admission to fellowship (she being a graduate of the University of Zurich). It was referred to a Committee, the majority of whom subsequently reported, that under the Acts 1789 and 1859,[1] Miss Dimmock was entitled to an examination for admission. A minority report objecting to this view, it was after discussion finally voted to recommit, and the Committee were instructed to ask legal advice. At the meeting in June, 1873, the opinion of Messrs. Hoar and Putnam was given, "that the Society had the power to admit or refuse to admit females to membership;"[2] upon which "it was voted to instruct the Censors not to admit females to examination as candidates for fellowship." And there, so far as this Society is concerned, the matter now rests.

From early times medical degrees have occasionally been conferred upon women, notably at Salernum[3] and Bologna, and it is claimed that they may

---

[1] Digest, Art. xx.

[2] For opinion, see Records, October, 1873.

[3] During the middle ages the celebrated school of Salernum produced several female physicians, and one called Trotula practised with great distinction and wrote a work on the diseases of women.—Tilt. Uterine Therapeutics, p. 2.

now graduate in Italy, France, Russia, Spain, Switzerland, and from a few medical schools in this country. With reference to the degrees conferred upon them at Bologna so often paraded, it is doubtful if "any woman ever received a degree there as the natural sequence of her studies without a special injunction of the Pope or Emperor." [1] In Scotland it has been recently decided that women, even if they can meet the required examinations, have no legal right to compel instruction from any medical school, and on their appeal from this decision to the House of Lords, the Professors of the University of Edinburgh claimed in their "Petition" that "the teaching of medicine in a University to a mixed class of young men and young women would be opposed to the convictions of a vast majority of the educated classes of the country." . No woman has thus far it appears ever studied and graduated in a Scotch, nor, so far as I can ascertain, an English University. At a convocation of the University of London in May, 1874, a motion for the admission of women to degrees was carried, 83 to 65. The meeting, which seems have been largely composed of women's supporters, drummed up for the occasion, "did not constitute more than a tenth of the whole number of members, few voting 'aye' belonging to the medical profession, and there is no reason to expect that the senate will consider themselves bound seriously to entertain the question." [2] In March, 1874, the London Obstetrical Society, after a full

---

[1] Saturday Review, July, 1873.
[2] Medical Times and Gazette, May 16, 1874.

discussion, declined to admit females to fellowship for the reason, among others, that "it will be neither for the good nor for the happiness of women that they should assume our habits, occupations and anxieties in addition to their own, which they cannot possibly throw off."[1]

It is a common popular fallacy founded upon a superficial, perverted view of the facts, that females are peculiarly adapted to the practice of Obstetrics. Nothing could be more erroneous, as there is probably no branch of the profession for which they . are so ill adapted. More than half a century ago the propriety of the employment of women as midwives seems to have agitated the profession and community in this vicinity. Among others the matter was reviewed by Dr. John Ware, whose opinion was clearly against it, not on the ground of any "intellectual inferiority or incompetency in the sex," but "rather from the nature of their moral qualities." He adds, "I venture to say that a female could scarce pass through the course of education requisite to prepare her as she ought to be prepared, for the practice of midwifery, without destroying those moral qualities of character which are essential to the office." His reasons are cogently put, and to the many here present who remember his wise judgment, the acuteness of his observation and the purity of his character, those reasons would have great weight.[2]

---

[1] London Obstetrical Journal, February, 1874.
[2] Remarks on Employment of Females as Midwives. Boston. 1820.

3

The first regularly educated practitioner who
devoted himself to obstetrics in this country, is
said to have been Dr. James Lloyd, of Boston,[1]
in 1754; and next after him, Dr. Shippen, of
Philadelphia,[2] in 1756. Previous to that time it
would seem that this important business was ex-
clusively in the hands of illiterate women, whose
practice seems to have been very little better
than that still in vogue among our western
savages.

A goodly proportion of the midwifery practice
of Great Britain among the poor and middling
classes is still performed by females, but the mor-
tality resulting from their ignorance is such that
the public are now crying for more stringent laws
for their governance—demanding urgently a better educated class and certain qualifications, the
want of which shall render the practice of their
vocation illegal.

A well-trained class of midwives would doubtless be cordially and universally welcomed by the
profession here. The poor everywhere are unable to render any adequate remuneration to the
educated and fully occupied physician, but can
afford a decent remuneration to midwives whose
preparation has been less costly, and whose expenses are comparatively small; but no such
woman should be permitted to practise until her
ability for the management of all simple cases has
been tested by a suitable examination. Such a

[1] Mass. Med. Society Communications, Vol. ii. 244.
[2] Tuner's Annals of Medical Progress, page 58.

class as the *sages-femmes* of the French and Germans would be invaluable here, sufficiently instructed to recognize impending danger, and under legal obligations to seek at once competent professional assistance in every abnormal case requiring interference.

Occasionally such exceptional women as Mesdames Boivin and Lachapelle have come to the surface, so skilled, so gifted, as to be recognized authorities even among their male confrères; but the very existence of such exceptional instances only adds strength to the argument against the general capacity of the sex for such work. Otherwise how do we account for the fact that their example has not been more generally followed? They were properly educated,—but the like facilities have always been equally open to others. Whence arises the significant fact that all this work has fallen so generally into the hands of the stronger sex? It must either be the result of an instinctive want of confidence on the part of the public, or because experience has proved that few women possess the masculine mental traits demanded in those grave emergencies which so constantly occur in the practice of midwifery.

In general practice these emergencies are seldom such as not to afford time for reflection, possibly a reference to authorities; but the occurrence of convulsions, post partum hæmorrhage, prolapse of the funis, placenta prævia, an inverted uterus, and the like, demand not only intelligence and the ready knowledge of what is required,

qualities which either sex may have in equal measure, but something more—the cool courage,
strength and prompt assumption of responsibility,
which few females with their quick feminine sympathies are equal to.

Indeed it may well be doubted if, in the whole
domain of even purely surgical practice, more of
these masculine qualities are required than in
some of the unforeseen accidents to which every
lying-in woman is liable, and upon the occurrence
of which nothing but the most prompt interference
will prevent a fatal result to the mother, or to the ·
child, possibly to both.   This branch of the profession—in the hands of women from time immemorial—receiving from them little or no improvement,
no thorough investigation, had no sooner passed
into the control of men than its practice was revo-
lutionized and so continuously developed by ex-
haustive study and analysis of the physiology,
mechanism and therapeutics of parturition, that it
may now be reckoned among the most perfect
departments of our art.

The complaint that educational facilities for
women have been wanting in our profession, is in
a large measure disproved by the instances already
cited.   Would it not be nearer the truth to say
that the number of females desiring to turn their
energies in this direction and to avail themselves of
such opportunities as they had, has been too limited to make any impression?   Were all restric-
tions removed as they desire, it is quite improbable that even now the applicants would perma-
nently increase.

Education favors the development of hidden faculties, but such faculties will always find their own methods for development, even if the best educational facilities are not within easy reach, and no amount of education will create faculties which have no latent existence. Even in our own sex the most brilliant examples of culture and success are by no means to be found among those who have been the most fortunate in their opportunities. True genius for any calling always finds a way to its desired end.

Does not all this failure on the part of women justify the apparent instinct which has led society to the belief, that the influence of sex in occupation is the natural sequence of the recognized sex in mind, and that—notwithstanding in every age there are certain disappointed natures, who, having either from lack of taste or opportunity missed their more appropriate vocations, endeavor as reformers to show that the fault is not in them —society has on the whole realized its best interests and acted accordingly?

To assert that women are not fitted for the training requisite for the mental development of man, implies no intellectual inferiority. "A superior woman may be injured by the mental training which is adapted to an inferior man," and vice versâ, which is only saying that the mental qualities of the two sexes have such inherent differences that no training is equally adapted to them indiscriminately. It does not follow that a higher system of general female education is to be discouraged ; it is only the methods to be pursued, and the ultimate aims to which it should be directed, upon which any difference of opinion should exist.[1] But admitting their intellectual equality, we may with perfect propriety question their physical ability. It has been claimed that this is " a matter for themselves, their parents and guardians " to decide ; but our reply to this is, that while to parents and guardians may be conceded the right to determine what they are willing to risk for their wards, it is no less true that there is a graver, broader issue, underlying such personal, partial ambition, and that the decision must, for the general interest of society, turn upon its relation to the public welfare, the healthy development of social life.

There are indisputably certain maternal, educational, domestic duties, the necessary result of original creative act, for which females alone are competent, and no part of which can be delegated to us, nor can there be any failure in their performance compatible with the social welfare of any community. If there is to be superadded to these essential and unavoidable claims upon female energy the study and practice of Law, Theology and Medicine, they should be endowed with something more than masculine, even superhuman powers.

---

[1] " The fact that the agitation for a better system of female education has been made a plea for the entrance of women into professions and vocations for which they are physically or morally unfitted, or from which they have been excluded by long tradition, has done much to injure a worthy cause."—*Lancet,* June 20, 1874.

It cannot be alleged of the medical profession, at least, that they are inimical to any movement tending to the better, more thorough education of women, in appropriate methods and directions; but they are not only entitled, it becomes their imperative duty, to raise a warning voice when they have reason to believe that such movements are taking directions opposed to the well-being of the race. In this view alone are the criticisms upon modern female education, which have excited such public attention of late, justifiable.

Whatever may be our opinion of the entire accuracy of the physiological statements which have been so confidently made, and although in the judgment of many of us this argument has been pushed much farther than the facts will warrant, it is very certain to most medical minds, that there is an underlying basis of truth justifying the feeling that some radical change is needed in the training of young women, if we would not have mothers not merely unable to nurse their babies, but if we would have any babies at all to require of them that impossible office.

Whether the difficulty be moral or mental, originating in violated physiological law, vicious social exigencies, or unavoidable climatic influences, the discussion, it is to be hoped, will result in arousing some effective action on the part of all who are interested in the happiness of coming generations.

The prevention of conception and the murders committed during the early months of pregnancy,

so notoriously common in these days, have more
to do with the nervous and uterine ill health of
women than they are willing to acknowledge,
or than their critics have seen fit to recognize.
Were every practitioner at liberty to make known
the facts in this direction, of which he has almost
daily cognizance, the public would be horrified.

The decided ground taken by this Society
against the lax morality prevailing as to this de-
struction of early fœtal life, is well known. No
abortionist can remain with us a moment longer
than is necessary for his legal expulsion, and there
is perhaps no other moral delinquency, the bare
suspicion of which in any one meets among us
with sterner reprobation.

In this connection, too, we must in fairness not
lose sight of the fact, with its evident bearing
upon the question at issue, that the subjects of
nervous exhaustion, hysteria, derangement of the
pelvic organs, and the whole train of disabilities
peculiar to the sex, are not confined to the so-
called better classes, nor even to that small portion of them supposed to be injured by excessive
brain-work.   Our hospitals and dispensaries,
whose occupants are mostly from the middle and
laboring classes, and who certainly cannot be sus-
pected of suffering from too great intellectual
activity, show quite as large a proportion.

Nor is it by any means true that the nervous,
delicate, thorough-bred American beauty, fragile
as she often seems, does produce a progeny really
inferior in either physical or intellectual strength,

to the coarser and rougher types prevailing in these other classes. Nothing in our social, certainly nothing in our political or military history, where only we can look for the test on a large scale, would justify such a supposition. It is asserted by most of those who have given attention to this subject, that the delicacy and sensitiveness of the female organization, and especially the physiological peculiarities of their sexual nature, are incompatible with the physical vigor required for the harassing and wearying duties of the medical profession, duties from which even the strongest men are often obliged to seek relaxation. To this the reply is made that the assertion is not founded upon fact. This answer may be, doubtless in some instances is, correct, but we may be permitted to question whether the cases cited in proof are not really exceptional, and whether such results, as a rule, can be attained without in some measure destroying that relationship between the sexes established by an allwise Providence, and the recognition of which has, heretofore, been thought essential to the best welfare of society. How many women after all are there, whose health, strength or temper for a fourth or fifth of the active period of their lives, is not accompanied by a state of nervous erethism sufficient to materially and unfavorably influence their mental equipoise, as well as their physical ability for professional work. "They cannot escape the physiological conditions of their sex." Most women as well as most men naturally hope to be married, and being married hope to have families. If their time is to be given to the exacting demands of professional life, instead of, or in connection with, those more domestic pursuits which have been heretofore considered as their appropriate sphere, and for which no one else is or can be made competent, they must first point out to us some feasible substitute, or both parties will come to grief.

There is no profession, probably few occupations heretofore reckoned as masculine, which do not require as we have said, for their successful prosecution, all of a man's energies, time and strength. Certainly the profession of medicine admits of no divided allegiance, and the physician who gives anything more than the time absolutely necessary for recreation, to other occupations, would be justly thought recreant to the public responsibilities which he has assumed.

On the other hand the woman, who from any cause has unfortunately—or as she may say, fortunately—failed to meet with the one who might have made an agreeable domestic life possible for her, may claim that she has no embarrassments to her freedom; but how many are there who will willingly admit, until the elasticity of youth and the best if not the only years fit for professional preparation have long departed, that she still does not look forward to the day when she shall have home cares and duties to absorb her time.

And here I cannot refrain from protesting against the unmanly and ungenerous sneers with which one critic after another has garnished his tale, when alluding to this so called "missed vocation." The argument obtains neither weight nor dignity from ridicule, and when clothed as it sometimes has been in language admitting only of indecent construction, no answer is possible from the other side. It has been well said that if women have "missed marriage," it is all the more creditable that they have "missed maternity;" to which we may add, that if men bore the babies, how many bachelors would escape conviction? Physiological considerations are an important, to many minds the most important, element in this discussion, forced upon us by medical women, and can only be met by unequivocal language; but surely our language is not so barren as to make coarseness essential to plain speaking. It is not found necessary in our Society discussions nor in general medical literature.

The right of women to attempt any sphere of intellectual development and activity is, as we have freely admitted, unquestionable; but an equal right pertains to the male sex—who from the beginning have governed, and probably will to the end continue to govern communities—to exercise their judgment also, in the practical decision as to what is wisest and best for the welfare and continuance of that aggregation of sexes which we style society, and the appropriate and natural combination of which alone makes civilization a possibility.

It has been eloquently asserted that the judgment of the world on the proper education and social status of woman has always been warped by the fact that men have made the laws and women have consequently never had an equal start in the race.[1] To show what women are capable of, where, for one reason or another,—the incubus being removed,—they have had free range for their abilities, the names are displayed, so familiar to every reader, of many noble women great in intellect, strong in will and clear in judgment, who have made their mark in history. Of these, so few when all told, a certain portion doubtless deserve all that is claimed for them, but of others equally great intellectually, it may be doubted whether they added anything to the sum total of human happiness in their day and generation. Granting all that has been claimed on this point, it is not at all proved that women as a whole have not had every opportunity which they have needed. Surely every advance made in the civilization of the human race has enured to their advantage and elevation as much as to ours. They from the beginning have undeniably had control of the training of the very young of both sexes, free from any restraining laws, and if they have taken such pains to develope the manly qualities of the boys and the feminine traits of the girls, it must have been either from an instinctive sense of their inherent differences, or because it has been soon discovered that a well-formed vigorous boy wont wear petticoats, nor a timid delicate girl pantaloons; although there is

---

[1] John Stuart Mill on "Subjection of Women," 1874, an authority which, however great it may be with those whose views he represents and whose cause he advocates, can hardly have much weight with others in view of his own peculiar relations with the sex.

occasionally some odd creature supposed to belong to the latter sex, whose infelicitous development in maturer years exhibits itself to the public in a hybrid costume indicative of an androgynous gender.

But admitting, if you please, for the occasion, that the medical profession does offer a proper opening for the ambition of a few of the female sex, the important consideration immediately intrudes itself, How shall that education be acquired ? Shall it be in mixed classes, in the companionship of the other sex, with all the existing advantages which our medical colleges render immediately available, with their organized corps of instructors, their dissecting rooms, laboratories, museums and cliniques? Or shall these same facilities be afforded them only in separate classes, involving of course double duty from the instructors, who for the present at least must be, for want of any other, of the male sex? Or, neither of these being acceptable, shall the privilege be accorded them of examination for degrees (as is now done in Dublin, Oxford and elsewhere) whenever they, by such independent methods as they best can secure, shall have fitted themselves for that agreeable ordeal? This latter method, it is needless to say, amounts practically to a requirement of the maintenance of separate facilities in the way of schools or colleges with all their expensive appurtenances, and of hospitals for the necessary clinical observation and experience.

Enough has already been said elsewhere as to the wide difference of opinion existing as to the propriety and safety of a general co-education of the sexes, and the discussion, though not always conducted in the wisest and best spirit, has undoubtedly caused a more correct appreciation of the real issues involved. Especially is this true in its application to the subject of medical education.

In Great Britain there seems to be a willingness to admit women, who may desire it, to matriculation, but only in separate classes. There being no law compelling unwilling professors to give separate instruction, their opportunities are necessarily limited. Even when, as stated above, they succeed in getting the necessary instruction, it has been formally decided "that women have at common law no right to demand to share the studies of men, at Universities, and no right to demand degrees." [1] Without registration, the practice of medicine is there illegal, and by the present law no one can be admitted to registration who does not possess a degree from some University. There the great preponderance of public sentiment seems decidedly adverse to any "mingling of the avocations assigned by custom from time immemorial to the different sexes," and especially to mixed instruction, as in itself unwise and impolitic.

On the Continent, however, there seems to be less sensitiveness, and women may be admitted freely, in mixed classes, to professional medical

[1] Saturday Review, July 5, 1873.

teachings and demonstrations on all subjects, without exciting disagreeable comment. Whether this be owing to more general freedom between the sexes, to what we should consider a lax tone of public delicacy, or to some other cause, it is difficult to say; but it is rather startling to suppose that such a state of public sentiment will ever become the rule in this country. Disclaiming any juvenile squeamishness, the idea of woman being present at certain anatomical demonstrations, either on the living or the dead subject, which are imperatively necessary for the proper instruction of students, is neither more nor less than disgusting. It is, in this connection, arrant nonsense to say that to the pure all things are pure, and to repeat the stupid platitudes about "prudery," "sickly sentimentalism," "false shame unworthy this advanced age," etc.; as well apply these terms to the ordinary conventionalities of all decently pure domestic life, and demolish every door and shield to privacy. Every feeling of refinement and delicacy instinctively revolts; and though the right to this freedom may be granted to those who are so anxious to assert it and who can face the ordeal, as we may equally admit their abstract right to do many things which would be disgusting to ordinarily decent and well-bred people, it may well be questioned whether this rude demolition of existing barriers is either judicious or desirable, especially for women themselves. The tendency must inevitably be to blunt that sensitiveness between the sexes which now forms one of the firmest holds of women upon men, and which, as a rule, ensures even to the weakest of the former the sympathy and support of every man who ranks himself above a brute.

It is said[1] that at Zurich women attend demonstrations of the male sexual parts, and the gynæcological cliniques with all that accompanying exposure which continental teaching permits and employs, and which among us is often so revolting even to men, without any apparent rudeness or objection from the students. This only proves the students to be wonderfully considerate and well-bred (which by the way is hardly accordant with their proverbial Bohemianism), or that the restraining influence of the Professor is exceptionally strong. The sensitive delicacy of the women themselves is sufficiently shown by their unblushing presence; but who that knows young medical students can doubt that such scenes in such companionship must, as already said, tend to deaden those feelings of respectful, tender and sympathizing appreciation which they of all persons most need to cultivate for their subsequent professional life and relations ?

On the other hand the difficulties attendant upon wholly separate instruction, with separate colleges, hospitals and other necessary paraphernalia, are so evident, that for any attempt in that direction disappointment may be predicted in advance. It seems hardly probable that the numbers seeking

---

[1] Tait, Medical Education of Women. Birmingham, 1874.

their advantages would be large enough to render
the solution of the pecuniary problem, that bug-
bear of all large educational projects, an easy one,
so that it apparently results in this, that if women
are to be educated as doctors with any prospect of
success, it must be effected through the instru-
mentalities already in operation for the instruction
of male students.

For many of the subjects required, some of the
graver objections to mixed classes do not exist.
Ophthalmology, Dermatology, Dentistry, Phar-
macy, Auscultation, Laryngoscopy, Chemistry,
Botany, possibly some others might be pursued in
common, obviating to that extent the need of double
duty from the instructors. But even in these there
would be the prerequisite of general anatomical
and physiological preparation, without which no
one should be permitted to graduate in any specialty. For their hospital instruction, also, there
would be no insuperable difficulty, as they could
be allotted separate days to follow the visiting staff,
few of whom would probably object to affording
the same facilities as are now accorded to male
students. But for most of the subjects, such as
Anatomy, Obstetrics, Theory and Practice and
Physiology, mixed classes ought to be decidedly
discouraged. A superficial teaching may doubtless
be given in such a way as not to be offensive, but
there are few instructors who would care to give the
elaborate demonstrations requisite for any really
thorough teaching; it would be sufficiently difficult
with a separate class of women unembarrassed by

the presence of a large number of young men. The expense, even when divided among the fullest classes, is always a burdensome matter to students, and for the limited number of females who would probably ever require this separate instruction, the burden would be proportionably increased,—a consideration however which does not concern us, nor indeed could they complain, as it would be infinitely less than the establishment of separate schools and hospitals.

Finally, granting that in one way or another such candidates have been able to prepare themselves sufficiently to pass the requisite examinations, and have received their degrees, the main question recurs—What should be the policy of this society as to admitting them to its privileges ?

If however much we may disapprove, they are to be professionally educated and given degrees, their admission seems to follow as a natural and necessary result, and indeed there are reasons why it might be thought advisable. It is quite improbable that in point of numbers their influence would ever become embarrassing.

It would terminate a so-called grievance, the constant iteration of which places us in a false position to the public, and renders our motives liable to misconstruction.

If women are to practise in the same field with ourselves, the recognized association would remove many awkward and anomalous positions between us, the public and themselves, inevitable so long as they are without the pale of the Society, and more than all would exercise a restraining influence by rendering them legally amenable to the ethics upon which we lay so much stress.

Still another reason consists in the fact that in the present position of these women society has no sure means of judging between the ordinary nurse who is called from her wash-tub to the assistance of her next-door neighbor in the pangs of labor—the woman who with impudence equalled only by her ignorance, offers for a thousand dollars to cure an ovarian tumor by some secret specific—the electrician and spiritualist, usually combined, who with closed eyes or in a heavenly trance, can with equal ease read the state of your mind or your liver—and that very limited number whom we now and then have reason to recognize as the educated and honest exceptions, and whom but for the disqualifications which we attribute to their sex we would willingly admit. The successful passing of our Censors' examination and subsequent admission to the Society, would make such a prominent distinction between these few latter and the whole festering mass of swindlers and abortionists as could not but be for the public good.

The most serious objection to their admission is, that it would be immediately construed as a tacit approval by the medical profession of any professional education for women. This would be a great error, the truth being that the profession as a whole are singularly unanimous in their disapproval of any such aim, they having a very decided conviction that the higher standard of

education which women are seeking, and which
they certainly ought to have, should find for its
development other and more appropriate spheres
which are as yet far from being exhausted.[1]

It is one thing to grant, as we have done, their
abstract right to any study or occupation which
does not contravene public law or morals, it is
quite another to be convinced of the necessity or
the propriety of many such occupations. Many
things, confessedly lawful, are manifestly inexpedient.

If, however, it be decided that our profession is
to be open to women, it is far better, that not only
their preparation for it but their action therein,
should be under competent supervision, and that
in neither should any unnecessary obstacle be
thrown in their way.

If, as we believe, they have undertaken a task

---

[1] The following remarks by Dr. John Lord, in a recent lecture on
Madame de Staël, are interesting in this connection : " He asked why
woman could not compete with man in any efforts that did not bring his
physical superiority into requisition? Common sense would tell us that
the sex could not be senators, jurists, or professors; not because their
intellects could not grasp great questions, but because the professions in-
curred labors which it was not becoming for a woman to assume. But
in those departments where labor was hid from the gaze or the intrusion
of the world, where public life was shunned, where lofty genius and
great attainments were required, why would not women not successfully
compete with men? Why should they not become linguists ; decipher
the difficulties of archæology; write the best poetry of the realm of sentiment ; be essayists and critics? Surely, they had lofty sentiments, were
keen to observe absurdities ; why should they not describe the life and
manners of former generations, if they had acuteness, patience, insight,
application? Why should they not be artists, if they had a quick sense
of the beautiful, the grand, the true? He knew of no more splendid future
for women than to encircle their brows with those proud laurels which
had ever been decreed to those who had advanced the interests of truth
and the dominion of the soul, and which experience showed they had
always won, and reason still more imperatively declared they might continue indefinitely to win."—*Reported in the Boston Advertiser*, April 16,
1875.

which will result in failure, we can well afford to
let it have a fair trial under new conditions, trust-
ing to the future to prove that which we had
thought already sufficiently proved by the experience of the past. Whatever may be the ultimate
decision of this Society, it is greatly to be desired
that it shall be influenced by no temporary or personal considerations, but that it shall turn upon
the broad, catholic ground of public and social
advantage. More than that, as integral parts of
the great whole we cannot ask ; less than that
will sooner or later create an antagonism on the
part of those from whom alone our rights are de-
rived, which will strip the Society of much of its
influence and power.

Another subject of vital importance to the public, to which I can here but briefly allude, is the
latitude now permitted by law to the operations
of unqualified practitioners, unqualified by training, unrecognized by any system or school, and
restrained by no other consideration than self-
interest. Every ignoramus who chooses so to do,
may with no other foundation than brazen impudence and a counterfeit name, assume the title of
doctor, spread broadcast his obscene literature and
advertisements over the fences and through the
mails, advertise in so-called respectable journals,
religious (sic) as well as secular, the vilest specific and the surest emmenagogue, with but the
most transparent veil to their real intent, and all
this without preventive interference from the

authorities, who apparently prefer to wait until
the morals are blunted, the money stolen, the in-
fanticide completed, after which they will punish
the offender if they can catch him.

The law recognizes no difference between such
irresponsible swindlers and those who, by many
years' devotion of their talents and means to the
acquisition of the knowledge which shall fit them
for the responsibilities involved in the duties of
the medical profession, have given, as it were,
securities in advance for their integrity and qualifications.

The public—and this in some measure it must
be confessed is our own fault, from the manner in
which we have dealt with it—are fully impressed
with the belief that our judgment of these charla-
tan dealers in patent nostrums is based upon self-
interest alone. To convince those who from want
of familiarity with it must necessarily take a very
superficial view of the subject, that all this irregu-
lar practice is pecuniarily a benefit rather than an
injury to us, will always be difficult if not impossible, until they are educated to appreciate the
exact truth—until they recognize, as we do, that
many ailments needing but a little judicious advice, with possibly some simple remedies for their
relief, are so aggravated by tampering with advertised nostrums until serious functional or even
organic changes are developed, as to require in the
end tedious and expensive attendance. Much of
the work of physicians, as we know, arises from
this neglect of proper early treatment, and more

still, perhaps, from those conditions of mind and body induced by the infamous and alarming statements as to the dangerous character of affections which are perfectly simple in their nature and treatment.

It has been argued that this is a subject which does not concern us, that such a state of things could not exist without the willing consent of the public, whose support and encouragement of them releases us from all responsibility. But is our responsibility so easily thrown aside? If, as is assumed, we, "encouraged by the patronage of the law," are endowed with certain privileges in order that we may the better diffuse a "knowledge of the animal economy and the properties and effects of medicines,"[1] it follows that we are under implied obligations to advise and educate, that in so doing we may, if possible, convince the community of any evil tendencies threatening their welfare.

The public at large not only consider it a right, but they would be the first to hold us derelict to our duty, did we not interfere in all hygienic matters, such as are involved in quarantine, drainage, the advent, causes and spread of infectious diseases, the investigation of the air they breathe and the water they drink. Indeed, it would seem that the only thing with which, in the view of certain minds, we must not meddle, is the liberty of any one to impose upon credulity with secret reme-

---

[1] See Charter of Society, Digest, Sec. I.

dies, which are by no means free from danger, as insidious but sure as the danger from any contagion or infection. It is no exaggeration to aver that the carminatives and soothing syrups, the balsams and elixirs, published as " perfectly harmless and warranted to contain no noxious ingredients," have slain their thousands.

The arguments used in opposition to legislation against quackery might as reasonably be applied to any legislation for other sanitary measures, authorizing the establishment of Boards of Health, which, notwithstanding the opposition met with at first, have now become fully established in public favor. The same right which not only justifies but demands of the State their interference with ventilation, drainage, food and drink, and the isolation of contagious diseases, exists and should be exercised also for the unrestricted sale of poisons and their congeners, secret remedies, as well as for the quasi practitioners who deal them out. The welfare of the people is the supreme law.

Few persons, of those at all conversant with the subject and who have rendered themselves competent to judge, will deny the need of some restraining influence in this matter; but what direction that influence shall take, what form it shall assume for the accomplishment of the end, is open to a wide difference of opinion. Perhaps the less we, as a profession, have to do with legislation the better, so long as our efforts in that direction are regarded with so much jealousy; but certainly the influence of the fourteen hundred medical men in

this State might, if exerted in concert and with
wisdom, so enlighten our community as to induce
them to demand for themselves better and more
efficient restraining laws, and, what would be more
to the purpose, obtain their cordial countenance
and support for the prosecution of these offenders,
as earnestly as they now require the trial, and on
conviction, the punishment of any other class of
criminals.

The claims in this direction of the public upon
us should henceforth be limited to the require-
ments of our charter — the proper diffusion of
sound medical knowledge in its broader and more
comprehensive sense. The responsibility for the
use or abuse of that knowledge should rest
with others. To be judge and jury at the same
time ought no longer to be required of us. As
the matter now stands, we are not only expected
to fight their battles single handed, but to pay the
onerous expenses, receiving as our reward no end
of abuse and misrepresentation, from the unreason-
ing prejudice that all this is done for the further-
ance of our own interests.

The attention of the profession, as the conserva-
tors of the public health, has been of late years
more and more awakened throughout the country
to the pernicious effects of the lawless liberty now
prevailing, and in some of the States this feeling
has culminated in legislation, giving supervisory
authority to local and State medical societies.

But any law which may by possibility, justly or
unjustly, be interpreted as for the benefit of. its

6

administrators, will never get the sympathy of the public, and lacking that must ultimately fail of its object.

If it could be made here, as in some countries, a penal offence to practise, or to dispense drugs, without a proper registration or license, obtainable only as the result of rigid but impartial examination by a competent board of examiners independent of the competition of any school or sect or society, all would probably be done that the nature of the case admits of, crushing out of sight, at all events, a host of impudent pretenders.

It is just here, however, that the main difficulty presents itself—that is, the appointment and composition of a board whose character and position should be so influential as to silence professional jealousies and compel the approval of the community for whose benefit alone their labors are designed.

Furthermore, if such a prohibitory law as is needed is to be anything more than a dead letter, the responsibility for its enforcement should rest with the examiners or some other independent but authoritative body. This Society might be active or not in bringing delinquents to the bar, but once there, the functions of the profession should be simply those of any other expert witnesses.

Mr. President and Gentlemen:

One duty remains to me, the only one which your favor has imposed that I approach with any feeling of reluctance; it is to remind you of the fact, that while we may be instrumental in warding off disease and death from others, we can ourselves claim no exemption from the dread summons.

During the past twelve months thirty-five of our number have gone to give an account of their stewardship. Most of them were long past the meridian of life ; two[1] highly distinguished for their varied acquirements, justified their title of Honorary Members by leaving behind them at three score years and ten an enviable professional and social reputation for scientific attainment and personal worth.

Among the others there were many whose long accustomed familiar presence at our meetings will be sorely missed. The perpetuation of their memory upon our records belongs of right to other hands, but I trust you will pardon me for an allusion to some who have been peculiarly prominent in this city.

There is hardly need of words to remind those of you, so long privileged with his daily presence, of our late President, Dr. CHARLES G. PUTNAM, a man who bore about with him such an invariable atmosphere of kindness and gentle courtesy, a man whose abundant professional resources, and whose great tact and operative skill in his own especial branch were always so freely at the service of his less experienced brethren, and whose qualities of head and heart it delighted us to re-

---

[1] Josiah Crosby, Manchester, N. H., died Jan. 7, 1875, æt. 81. Edward Delafield, New York, died Feb. 13, 1875, æt. 81.

cognize by conferring upon him our most honorable office.

The name and memory of JEFFRIES WYMAN, whose professional worth and high distinction in the more peculiarly scientific branches of our work, have already received such just and eloquent tributes, belong not to us alone, but to the whole country. To this Society it must always be a source of pleasure and pride that his name stands enrolled upon their catalogue as having been one of its most active members.

Dr. GEORGE DERBY, too, will be remembered and honored by all as one who, after achieving distinguished reputation for hospital service during the war of the Rebellion, deserved every credit for his persistent efforts in developing and sustaining the influence of our State Board of Health, until it finally became a permanent power for good in the land.

In like manner I might enumerate others of the list, were it not encroaching too much upon the privilege of those whose duty and pleasure it will be to commemorate them.

Let it be our aim so to conduct ourselves in our professional relations to each other and to the public, that we too at the last may deserve the like kindly recognition.

www.ingramcontent.com/pod-product-compliance
Lightning Source LLC
Chambersburg PA
CBHW022029190326
41519CB00010B/1636